NATIONAL AUDUBON SOCIETY® POCKET GUIDE

A Chanticleer Press Edition

Charles W. Chesterman
Honorary Curator of Mineralogy
California Academy of Sciences

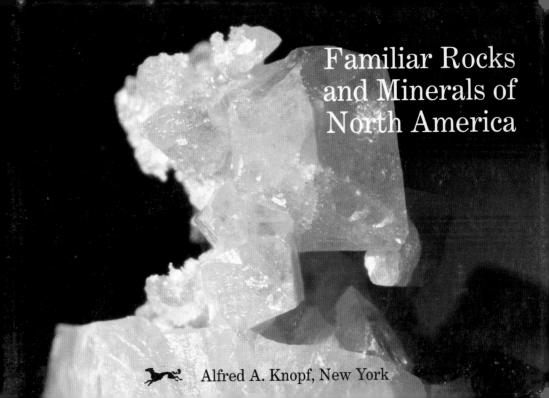

Familiar Rocks and Minerals of North America

Alfred A. Knopf, New York

Prepared and produced by Chanticleer Press, Inc., New
York. Color reproductions by Reprocolor International
s.r.l., Milan, Italy. Type set by Dix Type, Inc., Syracuse,
New York. Printed and bound by Toppan Printing Co.,
Ltd., Hong Kong.

Published March 1998
Ninth printing, March 2003

Library of Congress Catalog Card Number: 87-46020
ISBN: 0-394-75794-7

Contents

How to Use This Guide

Once considered a specialized hobby, the study of rocks and minerals is now a widely popular pursuit of immense interest and fascination to many people. This guide is for beginners; it presents easily discernible features that anyone can recognize and understand without having to use complicated keys or chemical tests.

Coverage This guide covers 68 minerals and 12 rocks, all commonly occurring in North America or frequently seen at shows. The geographical range spans from the Arctic through Mexico and from the Atlantic Coast to the Pacific.

Organization This easy-to-use pocket guide is divided into three parts: introductory essays; illustrated accounts of the rocks and minerals; and appendices.

Introduction The introductory essay "Distinguishing Rocks and Minerals" explains the basic composition of minerals and rocks, and how they are formed. "Identifying Minerals," illustrated with simple, clear charts, tells you what characteristics to look for in a mineral. "Identifying Rocks" gives practical clues for recognizing classes of rocks. "Collecting Rocks and Minerals" provides guidelines on where to look, the equipment needed, and what size specimens to keep.

The Rocks and Minerals This section includes 80 color plates, arranged visually by crystal shape, color, and overall appearance. Facing each illustration is a description of important field marks, including how to distinguish the rock or mineral from similar looking specimens. Also provided are its colors, its chemical composition, and a brief statement of its environment and where it occurs. For quick reference, a drawing of the most typical crystal form, habit, or aggregate shape supplements the identification description. An introductory paragraph presents additional facts about the mineral's appeal to collectors, its use, or the source of its name.

Appendices Following the species accounts, the appendices include a glossary of technical terms and an index.

Identifying rocks and minerals is not a simple matter, but once you learn the basics you will find a great deal of personal satisfaction. With a little practice, you will become familiar with the most common minerals and rocks, and discover that collecting them is an absorbing pastime.

Distinguishing Rocks and Minerals

Understanding the difference between rocks and minerals is easy—a rock is made up of minerals, usually several different kinds, while a mineral is homogeneous, that is, it has the same composition throughout. In a way, the two are like a patchwork quilt: the combination of patches in the quilt form our "rock," while each patch is equivalent to a "mineral."

What Is a Mineral?

A mineral is a naturally occurring chemical compound, with a homogeneous, recognizable composition. The molecules of every mineral are arranged in a particular crystalline structure that dictates the development of its crystal shape. The chemicals in a mineral account for its softness, shininess, or other noticeable properties. Some minerals have special ornamental value, and these are called gemstones.

What Is a Rock?

A rock is an aggregate of minerals: it may contain just one type of mineral, or it may be made up of many different sorts. Limestone is a rock composed largely of one mineral; granite pegmatite, on the other hand, may contain as many as 50 different minerals. Most rocks contain about six minerals.

The building blocks of the earth's crust, rocks are a vitally important source of information about our planet. The earth's history dates back more than four and a half billion years; rocks contain that history and reveal it to those who know how to read and interpret them.

Environment and Origin

The best clues to a mineral's origin lie in the rocks in which it occurs. Rocks are thus considered to be the mineral environment. Minerals found in igneous rocks (both plutonic and volcanic rocks) were formed by crystallizing in molten rock, called magma. The minerals in granite formed in an igneous plutonic environment, while those in basalt developed in an igneous volcanic environment.

Some minerals form by direct crystallization from a gas. For example, the mineral sulphur crystallizes from gas released at a volcanic vent. Other minerals, such as almandine and staurolite, develop deep in the earth's crust, where high pressure and elevated temperature metamorphosed sedimentary rocks. Read the essay "Identifying Rocks" to learn the different types of rocks. The essay "Identifying Minerals" explains basic mineral properties.

By knowing in which environments minerals occur, and which minerals are closely associated, you will be better equipped to know where to search for them.

Identifying Minerals

Minerals are recognized and identified by their distinctive, characteristic properties. Color is generally the most obvious, but since many minerals come in several colors, other properties, such as hardness, cleavage, fracture, the crystal form, habit of growth, cluster or aggregate form, as well as some other qualities, are all important. This guide mentions only those properties that can be verified by a beginner.

Hardness

The first property to examine is hardness, best verified on a fresh surface. In this guide, hardness is expressed as the ease with which a mineral can be scratched by a pocket knife. Minerals range greatly in hardness. Some, such as talc, are soft and can be scratched by a thumbnail; others, such as corundrum, can only be scratched by a diamond, the hardest of all known minerals.

Cleavage

Next, you need to examine the way a mineral breaks. Take a geologist's pick and strike your specimen with a single blow. If it breaks in two, forming one or more smooth, flat, lustrous surfaces, this kind of break is referred to as cleavage. Depending on the smoothness of the cleavage and the ease with which the mineral breaks, the cleavage is called *perfect*, *good*, *distinct*, or *poor*. Some species of minerals have more than one cleavage; others have none.

10

Fracture

If the mineral does not form a smooth, flat surface but is uneven, then this kind of break is referred to as the fracture. If the break produces a rough or somewhat lumpy surface, the fracture is *uneven*. If the surface forms a smooth curve like that of a seashell, it is called *curved*, or conchoidal. If the break creates a sharp, jagged surface like broken metal, it is considered *hackly*. Lastly, if elongated fibers or splinters are produced, the fracture is *splintery*.

Crystal Systems

Probably the most characteristic property of minerals is their crystal structure—the way in which the constituent chemicals are arranged. Because crystals always take a characteristic geometric form, most minerals can be identified as belonging to one of the six crystal systems, according to the way their atoms are arranged: isometric, tetragonal, hexagonal, orthorhombic, monoclinic, and triclinic. These systems describe the type and degree of symmetry that a crystal expresses. Isometric crystals have the simplest symmetry, triclinic the most complex. For many collectors, the beautiful and often elegant crystal forms are the most exciting aspect of mineral collecting. Because every specimen of the same mineral has the same crystal form, crystals are useful in identification.

11

Isometric Crystals Usually blocky, with many similar, symmetrical faces. They may be cubes, octahedrons (8-sided), dodecahedron (12-sided), or trapezohedrons (24-sided); crystals may be single or grow in combinations of one or more of these forms. Three axes of symmetry, all at right angles to each other and all of equal length.

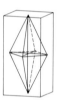

Tetragonal Crystals Generally long and slender or needlelike; characteristic forms are four-sided prisms, pyramids, double pyramids, tetrahedrons, and open forms called pinacoids, in which two opposing faces are the same. Three axes of symmetry: two axes, of equal length, lie in a plane at 90°; the third is longer or shorter and is at right angles to the others.

Hexagonal Crystals Usually prismatic or columnar, with hexagonal or rounded triangular cross section; characteristic forms are three- or six-sided prisms, pyramids, rhombohedrons, and pinacoids. Four axes of symmetry: three, of equal length, lie in a plane at 120°; the fourth axis is longer or shorter and is at right angles to the others.

Cube

Octohedron

Dodecahedron

Prism-pinacoid

Pyramids

Tetrahedron

Prism-pinacoid

Rhombohedron

Flattened pyramids-prism-pinacoid

Orthorhombic Crystals — Generally short and stubby, with diamond-shaped or rectangular cross section; characteristic forms are four-sided prisms, pyramids, and pinacoids. Three unequal axes, all at right angles to each other.

Monoclinic Crystals — Mostly stubby, with tilted matching faces at opposite ends forming a distorted rectangle; characteristic forms are prisms and pinacoids. Three unequal axes: two axes, at right angles to each other, lie in a plane; the third axis is inclined to the plane of the other two. There is one twofold axis.

Triclinic Crystals — Usually flattened, with sharp edges and sharp, thin cross section; no right angles on faces or edges; all forms are pinacoids. Three axes, all of different length and none perpendicular to the others.

Prism-pinacoid	Prisms-pinacoid	Flattened prisms-pinacoid

Prism-pinacoids	Prisms-pinacoid	Flattened prisms-pinacoid

Pinacoids	Twinned pinacoids	Pinacoids

Twinned Crystals

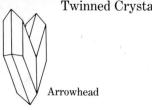

Arrowhead

When two or more crystals grow together in a definite manner they form twinned crystals. Contact twins are two simple crystals that grow together in the same plane, as in an arrowhead shape. In penetration twins, two separate crystals grow into a single structure; the corners of each crystal extend through the faces of the other, as in a right-angled cross, the Carlsbad twin, or a pseudohexagonal twin. Repeated twins are crystals that grow in parallel groups or occur at both ends of the crystal, as in side-to-side twins or knee-shaped twins.

Mineral Habits

Needlelike

Minerals take on characteristic forms called habits, that may not resemble the individual crystal but are its preferred way of crystallizing. Seven mineral habits are used in this guide: Needlelike habit refers to needle-shaped crystals, also called acicular. Bladed means broad and flat, like a knife blade. Branching depicts crystallizing in a treelike pattern, also called dendritic. Equant crystals have roughly the same diameter in every direction. Prismatic means elongated in one direction. Striated crystals have very shallow, parallel grooves on their faces. Tabular habit denotes flat, thin to thick crystals.

Right-angled cross

Carlsbad twin

Pseudohexagonal twin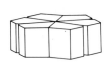

Side-to-side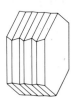

Knee-shaped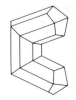

Bladed

Branching

Equant

Prismatic

Striated

Tabular

Botryoidal

Columnar

Coxcomb

Fibrous

Massive

Micalike

Radiating

Stalactitic

Mineral Aggregates   The minerals that we find are generally aggregates or combinations of imperfect crystals. There are eight common aggregated forms described here. Botryoidal aggregates resemble a cluster of grapes. Columnar forms grow in slender, parallel columns or prisms. Coxcomb aggregates show a serrated or comblike arrangement. Fibrous ones grow in threadlike fibers. Massive aggregates form interlocking grains, lacking apparent structure. Micalike forms grow in thin, flat, easily separated sheets or plates; also called micaceous. Radiating aggregates develop outward from a central point. Stalactitic forms resemble slender icicles.

Fluorescence   Another property useful in identification is the fluorescence of minerals. When exposed to ultraviolet light, some minerals fluoresce, or glow in various colors, including yellow, blue, green, orange, and red. Fluorescence can best be seen in a darkened room.

If you carefully examine each of these properties step by step, you should have enough information to identify the common minerals covered in this guide. Collecting minerals requires some patient observation of details but you will soon find this hobby both fun and rewarding.

Identifying Rocks

Although rocks are made up of minerals, rocks have some different characteristics from minerals and are therefore often identified by a different combination of properties. First, check the size of mineral grains and the texture of the surface.

Basic Minerals

Next, the most important step in identifying rocks is to learn to distinguish common minerals—especially quartz, the feldspars (orthoclase and microcline), biotite, muscovite, and hornblende. All of these basic minerals are covered in this guide.

Rock Types

If you have trouble identifying the minerals in a rock, you can still recognize the class to which a rock belongs. Rocks are classified into three broad types: igneous, sedimentary, and metamorphic.

Igneous

Igneous rocks form from molten rock, called magma, and are of two kinds—igneous volcanic and igneous plutonic. Volcanic rocks form directly from magma that cooled quickly on or near the earth's surface. Igneous volcanic rocks are generally fine-grained, have a smooth to rough surface, and sometimes small cavities. Basalt is a typical example of igneous volcanic rock.

Igneous plutonic rocks also form from magma, but it has

crystallized and cooled deep within the earth's crust. Slower cooling promoted the growth of mineral crystals, visible to the unaided eye. Plutonic rocks often appear speckled light and dark, like granite.

Sedimentary

The sedimentary rocks develop on the surface of the earth through the compaction and cementation of sediments that form when rocks decompose and disintegrate. Some sedimentary rocks are composed of grains of sand, as is sandstone. Others, such as limestone, develop on the bottom of oceans as the result of cementation of fragments of calcium-rich corals and seashells. A few sedimentary rocks are formed from chemically enriched waters: rock salt and rock gypsum are products of this process.

Metamorphic

The metamorphic rocks result from changes produced in preexisting igneous, sedimentary, or earlier metamorphic rocks by great pressure, temperature, or chemical activity within the earth's crust. Slate and marble are metamorphic rocks.

A word of caution: In many cases, it is only possible to make an approximate identification of a rock. Just to be able to distinguish between granite and sandstone is in itself an accomplishment.

21

Collecting Rocks and Minerals

You will find collecting rocks and minerals fun and satisfying. All you need are a pocket knife, a geologist's pick, a small cold chisel, and a magnifying glass, almost all of which can be obtained at a hardware store or hobby shop. A notebook is useful for recording information on the materials collected, and a supply of old newspaper is handy for wrapping and transporting your specimens.

Where to Look

Because rocks and minerals are almost everywhere, there are many places where you can search for good specimens. One of the best sources is the piles of waste rock called mine dumps, found at operating mines. Here specimens will be especially good if the mine is in a metalliferous mineral deposit, such as copper, gold, lead, silver, or zinc. At quarries where sand, gravel, and stone are being removed you can often find miscellaneous rocks from which individual crystals or clusters of crystals can be obtained. Although the sands along ocean and lake shores also contain minerals, bear in mind that the sand-size grains are usually too small to keep as specimens.

The mountainous regions of North America contain abundant exposures of various kinds of rocks. These are excellent places to search for rock and mineral specimens, with a geologist's pick and chisel in hand.

Clubs and Organizations Check with your local mineral society for schedules of field trips to nearby or even distant mineral localities. Mineral societies and clubs also have regularly scheduled meetings where members discuss the many fascinating aspects of their collections. Museums and college or university collections are another good source of information on places to collect minerals and rocks.

In the Field When out in the field, be sure to exercise care and good judgment: Wear goggles to protect your eyes from flying rock chips, and gloves when using tools. Never collect on private property without the owner's permission.

What to Save Beginners tend to collect too much material. Be selective in the numbers of specimens you save. Storage space will usually restrict the size of your specimens, as well as number. Rocks should be fist size in order to reveal the texture and mineral components but minerals can be smaller, even tiny micromounts—small crystals mounted in plastic boxes and viewed with either a hand lens or microscope.

Whether your collection is simple or elaborate, you will find collecting minerals and rocks will provide many hours of pleasure and relaxation.

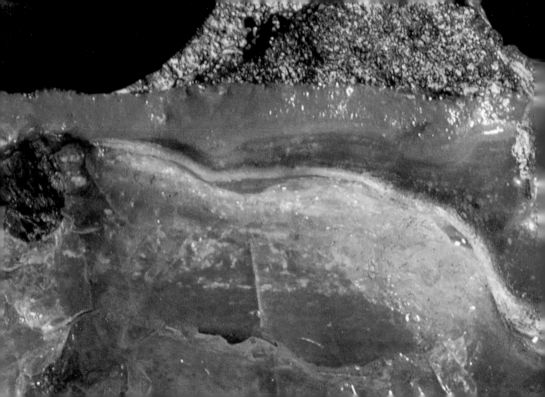

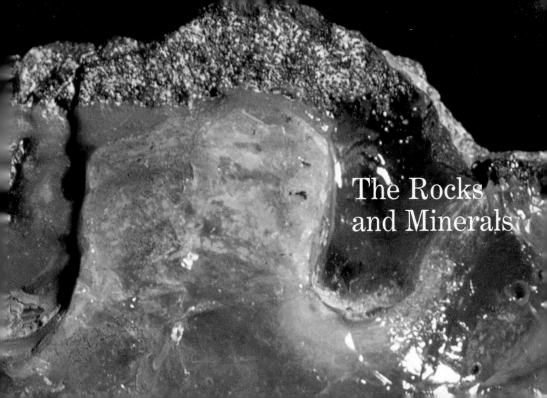

The Rocks
and Minerals

Phlogopite

The mineral phlogopite makes fine specimens; single crystals weighing as much as 90 tons have been known to occur. It is used to make insulation for electrical equipment. Like many related micas, phlogopite is asteriated—that is, when a small light source is viewed through a thin cleavage sheet, a star is visible. The name phlogopite comes from the Greek *phlogopos*, "fiery-looking," in allusion to this mineral's red-brown color.

Identification | Can be scratched by a knife; thin, micalike plates tough and very elastic; one perfect cleavage; translucent and transparent. Crystals monoclinic, often six-sided; asteriated. Biotite and muscovite are not asteriated.

Colors | Yellowish- or reddish-brown, gray to dark green; rarely colorless.

Environment | In marbles (metamorphosed limestone and rock dolomite).

Occurrence | Finest and largest crystals from Ontario; well-formed crystals from New Jersey and California.

Ilmenite

Often called titanic iron ore, ilmenite derives its name from a locality in the Ilmen Mountains of the USSR. It is an important ore of titanium and often contains much magnesium and manganese.

Identification   Cannot be scratched by a knife; brittle; weakly magnetic; lacks cleavage; curved fracture. Crystals hexagonal, commonly thick, tabular; also granular, compact. Magnetite is strongly magnetic; hematite has distinctive reddish-brown streak.

Colors   Iron-black, brownish-black.

Environment   Restricted to a few environments, principally in igneous plutonic and volcanic rock, often with magnetite in huge masses.

Occurrence   Fine crystals from deposits of emery in Massachusetts; with apatite and magnetite in Virginia; in deposits of magnetite in Wyoming; and with rutile, spinel, and biotite in Quebec.

Basic potassium, magnesium, iron, aluminum silicate

Biotite

Named after J. B. Biot, French physicist and mathematician, biotite is a mica. Like several other shiny yellow minerals, it is often called "fool's gold"; but despite the expectations it arouses, biotite has no value.

Identification Can be scratched by a knife; micalike plates tough and very plastic; one perfect cleavage; opaque to translucent. Crystals monoclinic, six-sided. Darker color distinguishes biotite from muscovite; phlogopite asteriated.

Colors Black, brownish-black, greenish-black, dark green, dark yellow.

Environment Develops in igneous plutonic and volcanic rock, and several types of metamorphic rocks (especially schists); common in granite pegmatites.

Occurrence No specific localities recorded in North America, but should be looked for in mountainous regions where igneous and metamorphic rocks abound.

Muscovite

Also known as isinglass and mica, muscovite is another member of the mica group. It is employed as a filler in industrial products, an insulator in electrical apparatus, and has been used to make windows for wood-burning stoves.

Identification Can be scratched by a knife; micalike plates flexible and elastic; one perfect cleavage; tough; transparent to translucent. Crystals monoclinic, often six-sided plates. Distinguished from biotite and many other micas on basis of color; phlogopite occurs in marble and is asteriated.

Colors White, colorless, yellowish, greenish, pink, brownish.

Environment Constituent of many igneous plutonic and metamorphic rocks; mined commercially in pegmatite bodies.

Occurrence Common, but good crystals up to 12″ across restricted mainly to granite pegmatites in Connecticut, Massachusetts, New Mexico, North Carolina, and Utah.

Molybdenite

A soft, platelike mineral, molybdenite is often mistaken for biotite, other micalike minerals, and graphite, which is actually a form of carbon. It is the important ore of molybdenum, an element used, like tungsten and chromium, to strengthen alloys. Molybdenum also plays a part in plant and animal metabolism.

Identification · Can be scratched by a knife; smooth surface; greasy, thin, micalike plates flexible; one perfect cleavage. Crystals hexagonal, forming six-sided plates; also scales, foliated masses. Flexible plates and weight distinguish molybdenite from biotite and phlogopite.

Colors Bluish-gray to lead-gray.

Environment Most common with quartz, chalcopyrite, scheelite, topaz, and fluorite in veins; in altered granites; and in metamorphosed impure limestones associated with scheelite, epidote, grossular, and chalcopyrite.

Occurrence Widespread and common; excellent specimens from California, Colorado, and Washington; also Mexico.

34

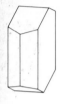

Silicate of calcium, magnesium, iron, aluminum

Hornblende

Like augite, hornblende is a member of a group of rock-forming minerals. It has no commercial value, but fine crystals enhance mineral collections.

Identification   Cannot be scratched by a knife; brittle; perfect cleavage in two directions, in shape of a diamond; fracture uneven to splintery; transparent to translucent. Crystals monoclinic, usually long to short prismatic; also fibrous, granular. Distinguished from tremolite and actinolite by color; tourmaline lacks cleavage.

Colors   Green, brown, black.

Environment   Common in igneous plutonic rocks (granite); in igneous volcanic rocks (basalt); and in schist and gneiss.

Occurrence   Fairly common, but fine crystals limited to a few locales, notably New Jersey and Ontario.

Pyrrhotite

Often called magnetic pyrite because of its magnetic properties, pyrrhotite usually contains small amounts of nickel and cobalt, and is in fact sometimes mined for these minerals. The name derives from the Greek *pyrrhotes*, meaning "redness."

Identification Can be scratched by a knife; brittle; uneven fracture; strongly to weakly magnetic. Crystals orthorhombic, commonly six-sided, but not hexagons; striated plates; also massive. Pyrite, marcasite, and arsenopyrite lack magnetism.

Colors Yellowish- to brownish-bronze; tarnishes to dark brown.

Environment Common in granite pegmatite associated with muscovite, microcline, orthoclase, and quartz; also associated with andradite, scheelite, and diopside in metamorphosed impure limestones; and with gold and galena in quartz veins.

Occurrence Well-known occurrences in California, Connecticut, Maine, Tennessee; also Ontario and Mexico.

Marcasite

Although appealing to collectors, marcasite disintegrates quickly, and should be kept in a dry place. Marcasite is sometimes called white iron pyrite, and in fact its name is believed to derive from an ancient Arabic word for pyrite.

Identification Cannot be scratched by a knife; brittle; distinct cleavage in two directions; uneven fracture. Crystals orthorhombic, commonly tabular; coxcomb aggregate. Pyrite is more common; whiter, softer, with different crystal shapes.

Colors Pale brass-yellow to almost white; tarnishes to deeper yellow.

Environment Occurs with galena and bornite in metamorphic rocks; with chalcopyrite and pyrite in vein deposits; and with galena, sphalerite, and dolomite in lead-zinc deposits.

Occurrence Widespread; some fine crystals from lead-zinc mines in Kansas, Missouri, Oklahoma, and Mexico.

Rhodochrosite

This important ore of manganese is used in steel-making. It often also contains iron, calcium, and zinc. The name "rhodochrosite" is from the Greek *rhodon*, "rose," and *chros*, "color."

Identification | Can be scratched by a knife; brittle; perfect cleavage in three directions, forming a rhombohedron; uneven fracture; somewhat transparent to translucent. Crystals hexagonal, commonly rhombohedrons; also cleavage masses, botryoidal, compact. Perfect cleavage in three directions distinguishes rhodochrosite from rhodonite, a harder manganese silicate with dark veins.

Colors | Pink, rose-red, dark red, brown.

Environment | Common in veins associated with chalcopyrite, galena, and sphalerite.

Occurrence | Finest North American rhodochrosite comes from mines in Colorado. Mines in Arkansas, Montana, and Newfoundland have produced massive pale pink rhodochrosite.

Dolomite

Named for Deodat de Dolomieu, a French mineralogist, dolomite is used to make magnesia, which is a component of antacids, refractory materials, and fertilizer.

Identification Can be scratched by a knife; brittle; perfect cleavage in three directions, forming a rhombohedron; curved, uneven fracture; transparent to translucent. Crystals hexagonal, usually six-sided rhombohedron; also granular, compact, massive. Calcite quite similar but can often be told by fluorescence, which is rare in dolomite.

Colors White, colorless, pink, gray, green, brown, black.

Environment In veins along with barite and fluorite and in metamorphosed limestone, also replaces calcite in limestones on the sea floor.

Occurrence Fairly widespread; finest crystals on the Ontario peninsula, between Lake Erie and Lake Huron; good pink crystals from lead-zinc ores in Mississippi Valley and in some gold-bearing veins in California.

Orthoclase

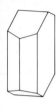

Long considered an important rock-forming mineral, orthoclase often occurs in beautiful crystals that eventually wind up in mineral collections and museums. Like microcline, orthoclase is one of the potash feldspars. It is a minor gemstone, used in ceramics.

Identification Cannot be scratched by a knife; brittle; good cleavage in two directions at right angles; uneven fracture; transparent to translucent. Crystals monoclinic, usually square to rectangular in cross section; tabular; twins are common. Right-angled cleavage and twinning usually distinguish orthoclase.

Colors White, pink, brown, gray, colorless.

Environment In many igneous plutonic, volcanic, and metamorphic rocks; also in granite pegmatites.

Occurrence Widely distributed throughout North America; excellent crystals from California, Colorado, Idaho, Nevada, New Mexico, and Virginia.

46

Halite

Known as rock salt, halite has many applications, but is chiefly used in the chemical industry as a source of sodium and chlorine, and to season foods. Large crystals are used to melt snow on roads and in churning homemade ice cream. The famous dry surface of Searles Lake in California is covered by a sheet of halite.

Identification   Can be scratched by a knife; brittle; tastes salty; perfect cleavage in three directions at right angles; curved fracture; transparent; fluorescent. Crystals isometric, commonly cubic with sunken faces.

Colors   Colorless or tinted with gray, yellow, red, or blue.

Environment   Widespread in dried ocean and salty lakes; often with gypsum deposits between layers of shale and sandstone.

Occurrence   Bedded halite deposits in Kansas, Michigan, and New York; also at Searles Lake, California.

Barite

This mineral is unusually heavy; its density makes it useful in plumbing the thick muds at oil-well drilling sites; it is also used in the manufacture of paints. Because it is opaque to X rays, barite has found significant medical applications.

Identification Can be scratched by a knife; brittle; perfect cleavage in one direction, good in a second, and distinct in a third; fracture uneven; transparent to translucent. Crystals orthorhombic, usually tabular, thin to thick; also compact, granular. Calcite not as heavy.

Colors White, gray, colorless; also various shades of yellow, brown, red, blue.

Environment Usually in veins containing metallic minerals, such as ores of copper, lead, and silver; also veins or masses associated with limestone.

Occurrence Widespread. Yellow tabular crystals in South Dakota and British Columbia; blue crystals in Colorado; colorless tabular crystals in California.

Albite

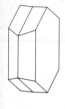

Rich in sodium, albite is one end of a chain of six closely related minerals that form the plagioclase feldspar group. Each mineral is closely related chemically to its neighbors. Pale opalescent albite, known as moonstone, is a prized gemstone.

Identification Cannot be scratched by a knife; brittle; good cleavage in two directions, poor in the third, with twinning striations on cleavage surfaces; uneven fracture; transparent to translucent. Crystals triclinic, common; usually flattened; twins show side-to-side arrangement. Orthoclase similar but lacks striations on cleavage surfaces.

Colors White, colorless, gray, opalescent.

Environment Often in important rock-forming minerals. Relatively pure albite found in igneous plutonic rocks, metamorphic rocks, and in granite pegmatites.

Occurrence Excellent albite crystals in pegmatites in California, Maine, and Virginia, and in schists in California.

Calcite

The most common of all carbonates, calcite is fun to discover, because clear crystals or fragments produce a double image of objects viewed through them. Also known as Iceland spar, calcite is used in optical instruments. It is the principal mineral in limestone.

Identification   Can be scratched by a knife; brittle; perfect cleavage in three directions, forming a rhombohedron; curved fracture (rare); transparent to translucent; fluorescent. Crystals hexagonal, showing six-sided rhombohedrons; may be granular, compact, stalactitic. Aragonite lacks rhombohedral cleavage; dolomite difficult to distinguish in the field.

Colors   White, colorless; also pale shades of gray, yellow, red, green, blue; brown to black when impure.

Environment   Chiefly in limestone; a minor constituent of many rocks; forms veins in places.

Occurrence   Widespread. Crystals up to 1,000 pounds found in New York; splendid crystals from lead-zinc mines in Kansas, Missouri, and Oklahoma; and in Mexico.

54

Celestite

Named from the Latin *caelestis*, "of the sky," in allusion to its sky-blue color, celestite is used in the manufacture of caustic soda and glass. Some of the best specimens in the world have come from North America.

Identification
Can be scratched by a knife; brittle; perfect cleavage in one direction, distinct in two other directions; uneven fracture; transparent to translucent. Crystals orthorhombic, usually thin to thick, tabular; also fibrous, nodular, in cleavable masses. Gypsum much softer.

Colors
White, colorless, bluish, reddish.

Environment
Formed in sedimentary rocks along with halite; also in veins with fluorite.

Occurrence
In numerous places in North America; especially in California, New York, on South Bass Island, Lake Erie; and in Mexico.

Colemanite

Discovered in Death Valley in 1887, colemanite was named for William T. Coleman, a prominent citizen of early San Francisco and a mine owner. This mineral makes attractive specimens; a source of boron, its comparatively low melting point makes it very useful as a flux in the manufacture of glass. It is easily identified because of its crystal shape and cleavage.

Identification Can be scratched by a knife; brittle; perfect cleavage, one direction, distinct in another direction; slightly curved fracture; transparent to translucent. Crystals monoclinic, commonly short prismatic, equant; also massive, compact, cleavable.

Colors Colorless, white, grayish- to yellowish-white.

Environment Sedimentary rocks along with other borate minerals.

Occurrence Only in W. United States, especially in Death Valley region. Nodules of colemanite with celestite occur in California and Nevada.

Gypsum

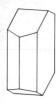

The mineral gypsum forms through the evaporation of mineral-enriched waters in lakes and small inland seas. The name is from the Greek *gypsos*, meaning "chalk." Gypsum is principally used in plaster and in the manufacture of portland cement. The colorless transparent variety of this mineral is called selenite; dense granular gypsum, or alabaster, is used in stone carving.

Identification Easily scratched by a knife; brittle; perfect cleavage in one direction, distinct in two others; curved and splintery fracture; transparent to translucent. Crystals monoclinic, commonly arrowhead-shaped twins; also granular, massive, fibrous, earthy. Muscovite and calcite similar but harder, not twinned.

Colors White, colorless, gray, yellow, red, brown.

Environment In layers associated with halite and limestone in former lake beds.

Occurrence Well-formed crystals from New York, Ohio, Utah, Mexico, and New Brunswick.

Kyanite

The name kyanite (formerly spelled cyanite) is derived from the Greek *kyanos*, meaning "dark blue." A minor gemstone, kyanite is used principally in ceramics. Although it is known in a variety of colors, it occurs most often as dark blue, and seldom in other colors.

Identification   Cannot be scratched by a knife; brittle; cleavage perfect lengthwise, good in second direction; splintery fracture; transparent to translucent. Crystals triclinic, individual crystals uncommon; usually in aggregates of bladed crystals. Tremolite and wollastonite white or gray.

Colors   Blue, colorless, green; less often white, gray; rarely black.

Environment   Product of metamorphism of clay-rich sedimentary rocks; often associated with biotite, quartz, and almandine in gneiss and schist.

Occurrence   Gem-quality blades have been found in North Carolina; known with rutile in Georgia; occurs in large masses in California.

Microcline

This is one of four potash feldspars, a group of important rock-forming minerals. A bluish-green to bright green form of microcline, called amazonite, is generally used by stone cutters to make beads, flats, and cabochons.

Identification Cannot be scratched by a knife; brittle; two good perpendicular cleavages; fracture uneven; transparent to translucent. Crystals triclinic, usually occurring as single crystals that show rectangular cross section; tabular. Excellent cleavage and bright color distinguish microcline from orthoclase.

Colors Bright green, bluish-green, white, pale yellow.

Environment Common in igneous plutonic rocks and granite pegmatites.

Occurrence Excellent crystals in California, Colorado, Connecticut, Maine, New Mexico, North Carolina, and Ontario.

Augite

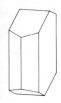

Named for the high luster of its crystals, augite comes from the Greek *augites*, meaning "brightness." This important rock-forming mineral has no particular economic value, but fine specimens usually enhance mineral collections, especially in museums.

Identification Cannot be scratched by a knife; brittle; two good, nearly perpendicular cleavages; fracture uneven; translucent. Crystals monoclinic, usually short prismatic; also columnar. Dark color distinguishes augite from diopside.

Colors Bright to dark green, grayish-green, brown, black.

Environment Igneous plutonic rocks and volcanic rocks (especially basalt).

Occurrence Excellent crystals from Colorado, New York, and Oregon; also Ontario.

Olivine

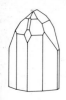

Named for its olive-green color, olivine belongs to a group of three closely related minerals; others are forsterite, which contains magnesium but no iron, and fayalite, which has iron but no magnesium. Combined, they form olivine, a source of the gemstone peridot.

Identification Cannot be scratched by a knife; brittle; cleavage indistinct, in two directions at right angles; curved and uneven fracture; transparent to translucent. Crystals orthorhombic, uncommon; usually occur as rounded grains; massive. Apatite similar but softer; tourmaline is striated.

Colors Yellowish-green (peridot), yellowish-brown, reddish.

Environment Typically in ultrabasic rocks and in volcanic basalt.

Occurrence Common in igneous volcanic basalt in Alaska, Arizona, California, Hawaii, New Mexico, Oregon, Washington, as well as British Columbia and Mexico; also in ultrabasic plutonic rock of California, North Carolina, and Washington.

Scheelite

Often scheelite has a distinctive bluish-white fluorescence that helps prospectors find the mineral. It is an important ore of tungsten, which is used in hardening steel and in electrical products.

Identification Can be scratched by a knife; brittle; curved fracture; cleavage distinct in one direction, poor in two others; transparent to translucent; flourescent. Crystals tetragonal, usually double pyramids, frequently tabular; also incrusting, granular, compact. Bluish-white fluorescence and heavy weight distinguish scheelite from fluorite.

Colors White, colorless, gray, yellowish, orange, brownish, greenish.

Environment In metamorphosed impure limestones with grossular, epidote, and molybdenite; in veins with quartz; and in granite pegmatites with beryl.

Occurrence Fairly widespread, especially in W. United States and British Columbia; noteworthy localities in Arizona, California, Idaho, Nevada, Utah, and Nova Scotia.

70

Quartz

Easily cut and polished, quartz is one of our most common minerals and a popular source of gemstones. It occurs in a very wide range of colors. Amethyst is the most valuable, but quartz is also used to make glass, oscillators, and filters for radios and telephones; and recently, quartz has replaced watch-springs.

Identification
Cannot be scratched by a knife; brittle; lacks cleavage; curved fracture; transparent to translucent. Crystals hexagonal, commonly six-sided prisms; also massive, granular. Beryl similar but harder.

Colors
Transparent, colorless (rock crystal); purple (amethyst); pink to rose-red (rose quartz); clear yellow (citrine); pale brown to black (smoky quartz).

Environment
Principal component of quartz veins; common as grains in many rocks, including granite, schist, gneiss, and sandstone; in beach and stream gravel deposits.

Occurrence
Widespread; rock crystals in Arkansas, California, and Ontario; smoky quartz in California and Colorado; amethyst in New Jersey; and rose quartz in California.

Topaz

Although gem-quality topaz comes in many colors, the red or purplish-red stones are artificially created by heat-treating brownish-yellow specimens. The name is from the Greek *topazos*, meaning "a precious stone."

Identification Cannot be scratched by a knife; brittle; one perfect cleavage; fracture slightly curved to uneven; transparent to translucent. Crystals orthorhombic, usually stubby to long prismatic and striated lengthwise; also granular, massive. Hardness and cleavage distinguish topaz from quartz and beryl.

Colors White, colorless, yellow, brownish-yellow (sherry-colored), pink, bluish, greenish, orange (hyacinth).

Environment Igneous plutonic and volcanic rocks; also in several kinds of metamorphic rocks, and in granite pegmatites.

Occurrence Pale blue and sherry-colored crystals from New Hampshire and Colorado; well-formed colorless and sherry-colored crystals in volcanic rock in Utah; gem-quality topaz from S. California; yellow topaz in Mexico.

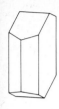

Calcium magnesium silicate

Diopside

This is an important rock-forming mineral of the pyroxene group. Crystals often occur with two sets of prism faces. Although diopside has no commercial value, rare high-quality material is a source of gemstones.

Identification — Cannot be scratched by a knife; good cleavages in two directions lengthwise, nearly at right angles; uneven fracture; transparent to translucent. Crystals monoclinic, usually short prismatic with good terminations; also massive, granular. Distinguished from olivine and epidote on basis of cleavage.

Colors — White, colorless, grayish, greenish, pink.

Environment — Common in certain metamorphic rocks, especially those derived from impure limestones, where it commonly occurs with dolomite, calcite, fluorite, grossular, phlogopite, and tremolite. Rare in igneous rocks.

Occurrence — Fine crystals from Ontario, large specimens to 6″ from California; New York and Quebec.

Rutile

This mineral often occurs in a more complex form, containing substantial amounts of iron. The name rutile derives from the Latin *rutilus*, "reddish," in allusion to its color. This is the principal ore of titanium; rutile is also used as a source of gemstones and in the manufacture of paints and ceramic glazes.

Identification Cannot be scratched by a knife; brittle; cleavage good in two directions, poor in a third; uneven fracture; translucent to transparent. Crystals tetragonal, commonly long striated prisms; also knee-shaped twins; granular, compact. Hematite also red to reddish-brown, but does not occur in striated crystals or in twins.

Colors Red, reddish-brown, black.

Environment Most often found in several kinds of metamorphic rocks such as schists; also in plutonic rocks.

Occurrence Good crystals in Arkansas, California, and Georgia; also common in beach sands in Florida and South Carolina.

Aragonite

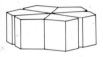

Named for the province of Aragon, Spain, where it was first found, aragonite is valued only as a specimen for collectors. This mineral is fluorescent; it gives off a red or yellow glow when exposed to ultraviolet light.

Identification
Can be scratched by a knife; brittle; good cleavage in one direction, poor in two other directions; slightly curved fracture; transparent to translucent; fluorescent. Crystals orthorhombic, usually short to long prisms, often twins in pseudohexagonal form; also fibrous, needlelike, columnar. Twin crystals and long cleavage fragments distinguish aragonite from calcite.

Colors
White, gray, colorless, yellow, green, violet, brown.

Environment
With gypsum and calcite in sedimentary rocks; in metamorphic rocks; and in copper mineral deposits with azurite and chalcopyrite.

Occurrence
Comparatively rare in North America; excellent crystals and hexagonal tablets from New Mexico; fine stalactitic masses from Arizona.

Pyromorphite

The name pyromorphite derives from the Greek words for "fire" and "form"; this mineral is so named because a melted globule assumes a crystalline shape upon cooling. Along with apatite, pyromorphite is a fairly common mineral in the phosphate group; all of these minerals form through the alteration of other minerals. Pyromorphite is a minor ore of lead and may contain traces of arsenic. The color and the hollow-looking crystals are good field marks.

Identification Can be scratched by a knife; brittle; lacks cleavage; fracture slightly curved to uneven; translucent. Crystals hexagonal; six-sided prisms common and often cavernous, or hollow at the tip.

Colors Green, yellow, brown, white, gray.

Environment A product of an alteration in lead-bearing sulfide minerals; often associated with barite.

Occurrence Occurs in Idaho, in the steel-mining district of British Columbia, and in Mexico.

Tremolite

This mineral is named for the site of its discovery—Val Tremolo in the Swiss Alps. Translucent, very fine-grained tremolite is familiarly known as nephrite jade. The rulers of ancient China had a passion for this material, and for thousands of years it has been fashioned into jewelry, ornaments, and sculpture.

Identification Cannot be scratched by a knife; small fibers flexible; two perfect cleavages in shape of diamond; uneven fracture; transparent to translucent. Crystals monoclinic, commonly fibrous; also radiating, bladed. Cleavage distinguishes tremolite from wollastonite.

Colors White to gray, yellowish, greenish, pink, colorless.

Environment Common in metamorphosed impure limestones; associated with calcite, grossular, diopside, and phlogopite.

Occurrence Fairly widespread; fine crystals from Ontario, pink crystals from New York, greenish crystals from Connecticut.

Corundum

Easily polished, corundum is a source of gemstones; deep red corundum, ruby, is extremely rare; blue corundum is sapphire. Where it occurs with magnetite, hematite, or spinel, corundum is called emery. This abrasive is used for sandpaper and nail files.

Identification | Cannot be scratched by a knife; brittle; lacks cleavage, but has parting in three directions; uneven, curved fracture; transparent to translucent; fluorescent. Crystals hexagonal, tapering, six-sided prisms; also tabular, striated. Striations on parting surfaces and hardness distinguish corundum from albite.

Colors | White, gray, brown to black (emery), deep red (ruby), blue (sapphire).

Environment | Several distinct environments: with albite in granite pegmatites; and with muscovite, almandine, and quartz in gneisses and schists.

Occurrence | Gem-quality corundum from Montana and North Carolina; excellent crystals in Canada, Montana, and Pennsylvania; also found in California.

Andalusite

Named for the Spanish province of Andalusia, this mineral is used in ceramics—notably for spark plug insulation. Andalusite is only secondarily a source of gemstones.

Identification Cannot be scratched by a knife; two good cleavages, nearly at right angles; uneven to slightly curved fracture; transparent to translucent. Crystals orthorhombic, usually stubby prisms with square cross section (chiastolite); also columnar, granular. Distinguished from tourmaline by square cross section; good cleavage.

Colors White, gray, pink, reddish-brown, olive-green.

Environment Andalusite develops with quartz and muscovite in granite pegmatites, and with biotite, corundum, almandine, muscovite, and topaz in metamorphic rocks.

Occurrence Massive pinkish-gray and white andalusite from California, and fine crystals from California and Massachusetts.

Staurolite

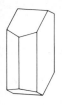

Often staurolite occurs in twin prismatic crystals that meet at right angles to make a cross (*stauros* in Greek); this form is known as penetration twinning. Transparent staurolite can be cut and polished and is used as a gemstone.

Identification Cannot be scratched by a knife; brittle; one poor cleavage; fracture uneven to slightly curved; translucent to opaque. Crystals monoclinic, commonly twins in right-angled cross. Andalusite and tourmaline similar but do not occur as twins.

Colors Yellowish-brown, reddish- to brownish-black.

Environment In schists; usually associated with albite, biotite, and quartz.

Occurrence Beautiful twin and single crystals occur in California, Georgia, Maine, New Hampshire, and New Mexico.

Titanite

This mineral gets its name from its high titanium content. Its alternate name "sphene" derives from its characteristic wedge-shaped crystals—"sphene" means "wedge" in Greek. Beautiful yellow titanite is a source of gemstones. Large crystals, weighing up to 80 pounds, have been found in Ontario and New York.

Identification Cannot be scratched by a knife; brittle; distinct cleavage in two directions; curved fracture; transparent to translucent. Crystals monoclinic, usually wedge-shaped. Staurolite similar but with poor cleavage in one direction and usually twinned crystals; sphalerite is softer.

Colors Brown to black; yellow, gray, green.

Environment A minor accessory mineral in many igneous plutonic and metamorphic rocks; also in granite pegmatites.

Occurrence Large crystals from Ontario and New York; gemlike crystals from New York; fine brown crystals from Montana; and vivid green crystals from Mexico.

Basic calcium aluminum silicate   **Vesuvianite**

This mineral goes by many different names—vesuvianite for Mt. Vesuvius, idocrase, Greek for "changing forms," and californite or California jade because of its jadelike characteristics. It is a source of gemstones.

Identification Cannot be scratched by a knife; brittle; poor cleavage in one direction; uneven to curved fracture; transparent to translucent. Crystals tetragonal; usually short prisms, square in cross section; commonly massive, compact. Distinguished from green garnet (andradite) on basis of superior hardness and crystals.

Colors Brown, green; rarely yellow or blue.

Environment Mostly in metamorphic rocks; massive variety— "californite"—occurs in serpentinite. Nicely shaped crystals common in metamorphosed impure limestones and associated with grossular, calcite, and diopside.

Occurrence Gem-quality crystals in Quebec; 2″ crystals in California and Montana; massive vesuvianite in N. California; yellow prismatic crystals in Mexico.

94

Basic silicate of calcium, iron, and aluminum

Epidote

Often called pistacite because of its characteristic pistachio-green color, epidote is both a mineral and name of a group of related minerals that includes zoisite, clinozoisite, piemontite, and allanite. Epidote is occasionally used as a gemstone.

Identification Cannot be scratched by a knife; brittle; perfect cleavage in one direction (lengthwise); fracture uneven; transparent to translucent. Crystals monoclinic, usually long, slender, grooved, prismatic. Actinolite similar but lacks grooved crystals, different colors.

Colors Yellowish-green to brownish-black.

Environment Usually metamorphic, forming with metamorphosed limestone; often associated with grossular, scheelite, and molybdenite. Also in some granite pegmatites and schists.

Occurrence Excellent examples from Alaska, California, Colorado, Nevada, and Mexico.

Beryl

This one mineral is the source of several gemstones, including emeralds and aquamarine. It is also the principal source of beryllium, an element used as a hardening agent in certain alloys.

Identification   Cannot be scratched by a knife; brittle; indistinct cleavage, one direction; fracture uneven to curved; transparent to translucent. Crystals hexagonal, usually prisms, striated lengthwise. Quartz and apatite are similar but softer, with crosswise striations.

Colors   Bright green (emerald), greenish-blue (aquamarine), yellow (golden beryl), red, pink (morganite), white, colorless.

Environment   Most common in granite pegmatites, associated with quartz, muscovite, and microcline; also in schists associated with muscovite, quartz, and almandine.

Occurrence   With granite pegmatites in S. California, Colorado, New England, and North Carolina; occurs with scheelite in granite pegmatite in Nevada.

Calcium fluorine-chlorine-hydroxyl phosphate

Apatite

Occurring in many colors, apatite is easily mistaken for fluorite, olivine, and aquamarine, the greenish-blue form of beryl. Its name comes from a Greek word meaning "to beguile." The most common phosphate-bearing mineral, apatite is used to make fertilizers.

Identification Cannot be scratched by a knife; brittle; cleavage poor, in one direction; fracture curved and uneven; transparent to translucent. Crystals hexagonal, commonly prismatic; also tabular, granular, or massive. Tourmaline and quartz lack striations, have different crystal shapes.

Colors Green, brown, red, yellow, violet, pink, white, colorless.

Environment Many igneous and metamorphic rocks; especially abundant in granite pegmatites and in rocks formed through metamorphism of impure limestones.

Occurrence Large crystals in marble deposits in Ontario; violet to blue, stubby crystals in Maine; green crystals in S. California; yellow crystals with magnetite in Mexico.

Tourmaline

This mineral's name means "mixed colored stones" in Sinhalese, referring to varicolored pebbles found in sand and gravel. It is a source of gemstones.

Identification Cannot be scratched by a knife; brittle; lacks cleavage; uneven to curved fracture; transparent to opaque. Crystals hexagonal, with a rounded triangular cross section and lengthwise striations; also radiating, columnar. Distinguished from apatite on basis of superior hardness; beryl is not striated.

Colors Black (schorl), blue (indicolite), pink and red (rubellite), brown (dravite), green, multicolored; rarely white.

Environment In granite pegmatites and other plutonic rocks, and metamorphic rocks; rare in veins. Rubellite in lithium-rich granite pegmatites; schorl in simple granite pegmatites; dravite in complex granite pegmatites.

Occurrence Excellent and gem-quality tourmaline from California, Colorado, and New England.

Chromite

First discovered in 1797, chromite is the principal ore of chromium, an element used in making alloys and in electroplating. A kind of garnet called uvarovite is commonly found with chromite. The mineral's name comes from the Greek *chroma*, meaning "color," because the chromium compound imparts strong coloring that can be used as pigment.

Identification Cannot be scratched by a knife; brittle; no cleavage; curved fracture; may be slightly magnetic. Crystals isometric, uncommon; eight-sided, usually granular, compact. Magnetite similar but strongly magnetic.

Colors Iron-black, brownish-black.

Environment In serpentinite and ultrabasic rocks; also in sand and gravel deposits formed from these rocks.

Occurrence Maryland and Pennsylvania; mined extensively in California, North Carolina, Oregon, and Wyoming.

Andradite

This member of the garnet group is named for J. B. de Andrade e Silva, Brazilian geologist. The mineral is a source of gemstones and a minor source of abrasives; a very dramatic, brilliant, all-black form is known from a few localities.

Identification Cannot be scratched by a knife; brittle; lacks cleavage; curved fracture; transparent to opaque. Crystals isometric, commonly 12-sided; also granular, massive. Green, yellow, and black andradite easily told from other garnets (almandine, grossular); crystal shape distinguishes it from tourmaline.

Colors Olive-green (demantoid), yellow to brownish or bronze-green (topazolite), black (melanite).

Environment In granite pegmatites with albite and biotite; also with epidote, magnetite, and scheelite in metamorphosed impure limestones.

Occurrence Melanite occurs in California and New Jersey; demantoid in California and Pennsylvania; topazolite in Arizona.

Grossular

A member of the garnet group of minerals, which includes pyrope, almandine, spessartine, andradite, and uvarovite, grossular is perhaps the commonest of all six garnets. It is easily cut and polished, and a source of gemstones as well as an industrial abrasive.

Identification Cannot be scratched by a knife; brittle; lacks cleavage; curved fracture; transparent to opaque. Crystals isometric, usually 12-sided or 24-sided. Crystal form and hardness distinguish it from apatite and tourmaline.

Colors Colorless, white, yellow, pink, green, brown.

Environment Metamorphosed impure limestones; commonly with wollastonite, calcite, epidote, and vesuvianite.

Occurrence Most significant North American occurrences are in California, Maine, and Quebec; fine white crystals, up to 4″ across, from Mexico.

Copper

Since earliest times, humans have known how to mine this mineral. It is "native"; that is, copper—like silver, gold, and a few other minerals—occurs in a relatively pure and usable form in nature. The word copper comes from the Greek *Kyprios*, the island of Cyprus, where copper was early found and used.

Identification	Can be scratched by a knife; ductile; malleable; lacks cleavage; fracture hackly. Crystals isometric, usually cubic and 12-sided; also scales, lumps; branching. Malleability and color distinguish copper from gold and silver.
Colors	Copper-red turning black, blue, or green with tarnish.
Environment	Frequently found in metamorphosed volcanic basalt associated with silver and zeolites, a group of hydrated aluminosilicates that can lose part or all of their water without changing crystal structure.
Occurrence	Masses of native copper found in Michigan and Ontario; other sources in Alaska, Arizona, New Jersey, New Mexico, Nova Scotia, and Mexico.

Galena

The most important ore of lead, galena became a lynchpin in the early home-entertainment industry: It was the crystal in crystal-detector radios, bringing Tom Mix and others into thousands of living rooms.

Identification Can be scratched by a knife; brittle; perfect cleavage in three directions at right angles; slightly curved fracture. Crystals isometric, usually cubic or cubes combined with 8-sided or 12-sided figures; also massive, fibrous. Cubic cleavage and lead-gray color distinguish galena from sphalerite.

Colors Dark lead-gray.

Environment Common; several modes of occurrence. Associated with andradite, pyrite, and chalcopyrite in certain metamorphic rocks; with quartz, bornite, and chalcopyrite in veins; and with marcasite and sphalerite in lead-zinc ore deposits.

Occurrence Fine specimens from California, Colorado, Idaho, Kansas, Missouri, and Oklahoma.

Fluorite

Although it occurs in crystals, fluorite is too soft to be a good gemstone. It is used to promote fusion in the smelting of metallic ores and in the manufacture of steel and hydrofluoric acid.

Identification Can be scratched by a knife; brittle; perfect cleavage in four directions, forming eight-sided cleavage fragments; uneven fracture; transparent to translucent; fluorescent. Crystals isometric, usually cubic, rarely eight-sided; also cleavage masses, compact, granular. Calcite and quartz similar but lack perfect eight-sided cleavage and crystal shape.

Colors Violet, blue, green, yellow, brown, bluish-black, pink, rose-red, colorless, white.

Environment Associated with albite, calcite, and pyrite in metamorphic rocks; with rhodochrosite and celestite in veins.

Occurrence Most important occurrences in Colorado, Illinois, Kentucky, Ohio, Ontario, and Mexico.

Spinel

Red spinel is often misidentified as ruby. In fact, its appearance is so convincing—it is ruby-red in color and contains few flaws—that it was commonly substituted for ruby in the crowns of England and Russia. Thus its use as a gemstone is well established.

Identification
Cannot be scratched by a knife; lacks cleavage; curved fracture; transparent to opaque. Crystals isometric, commonly eight-sided. Crystal shape and hardness distinguish spinel from the garnets.

Colors
Red, green, blue, black, brown.

Environment
In metamorphosed impure limestones; usually associated with vesuvianite, phlogopite, grossular, and calcite.

Occurrence
Sharp spinel crystals in California, New York, New Jersey, and Ontario.

Almandine

Sometimes mistaken for red spinel or ruby, almandine is a member of the garnet group, of which there are six distinct but closely related members. Almandine is a source of gemstones and, like other members of the group, is also used as an industrial abrasive. Unlike the related pyrope and rhodolite, almandine is likely to contain flaws.

Identification | Cannot be scratched by a knife; brittle; lacks cleavage; curved fracture; transparent to opaque. Crystals isometric, usually 12-sided or 24-sided. Crystal form and color distinguish almandine from other garnets and tourmaline.

Colors | Deep red to brown or brownish-black.

Environment | Most common in metamorphic rocks derived from aluminum-rich sedimentary rocks; often associated with biotite, hornblende, and andalusite in schists. Rarely in igneous plutonic rocks.

Occurrence | Gem-quality crystals occur in schist in Alaska, Idaho, and Michigan; also in granitic rock in S. California.

Magnetite

Sometimes called lodestone, magnetite is strongly magnetic and will attract small pieces of iron and steel. A common mineral, its principal use is as an ore of iron. Because magnetite is resistant to rock alteration, grains tend to accumulate in sands, where the mineral can easily be detected by a magnet thrust into the sands of beaches and streams.

Identification Cannot be scratched by a knife; brittle; no cleavage; fracture slightly curved, uneven; strongly magnetic. Crystals isometric, usually eight-sided; may be massive, granular. Magnetism usually distinguishes magnetite from chromite.

Colors Iron-black.

Environment Common; a minor constituent in many igneous, sedimentary, and metamorphic rocks.

Occurrence Found almost anywhere; fine crystals from California, New Mexico, New York, Utah, Vermont, Nova Scotia, and Ontario.

Sphalerite

The commonest of all zinc-bearing minerals, sphalerite is often called zincblende. It frequently contains iron, which imparts a dark color to the mineral. Sphalerite is the principal ore of zinc. The name derives from a Greek word meaning "treacherous," in allusion to its similarity to other minerals, with which it used to be confused.

Identification Can be scratched by a knife; brittle; perfect cleavage in six directions; transparent to translucent; occasionally fluorescent. Crystals isometric, 8-sided or 12-sided; also granular, compact. Distinguished from galena on basis of cleavage and color.

Colors Yellow, brown, red, green, black; rarely white or pale gray.

Environment Often with galena in several kinds of metamorphic rocks, veins, and other lead-zinc mineral deposits.

Occurrence Outstanding specimens from mines in Colorado, Kansas, Missouri, New Jersey, Oklahoma, and Mexico.

Bornite

Also called peacock copper because of its varicolored appearance, bornite is named for Ignaz von Born, a famous Austrian mineralogist. Bornite is an important ore of copper. Although very fine crystals are seldom encountered, there are many localities where good examples of massive bornite can be found.

Identification Can be scratched by a knife; brittle; lacks cleavage; uneven to curved fracture. Crystals isometric, uncommon; cubes, 12-sided forms more common; compact, granular, massive. Chalcopyrite, pyrrhotite, and pyrite similar, but lack multicolors, are somewhat harder.

Colors Copper-red to bronze-brown with deep blue tarnish creating varicolored effect.

Environment Often in metamorphic rocks associated with andradite, pyrite, barite, and calcite; also in quartz veins associated with chalcopyrite and pyrite.

Occurrence Widespread in mining areas of Alaska, Arizona, California, Colorado, Connecticut, British Columbia, and Quebec.

Chalcopyrite

Frequently referred to as "fool's gold" because of its yellow color, this abundant mineral is named from the Greek words *chalkos*, "copper," and *pyrites*, "fiery." It is an important ore of copper, and its presence often indicates that malachite is nearby.

Identification Can be scratched by a knife; brittle; one poor cleavage; uneven fracture. Crystals tetragonal, usually four-sided; also compact, massive, granular. Softer and more brittle than pyrite; gold is malleable.

Colors Brass-yellow, golden-yellow; often tarnishes iridescent purple or black.

Environment Many environments: with pyrite, barite, andradite, and molybdenite in several metamorphic rocks; with pyrite and gold in quartz veins; and with galena, sphalerite, and dolomite in lead-zinc mineral deposits.

Occurrence Widespread; excellent examples from Arizona, Kansas, Missouri, Oklahoma, Pennsylvania, Utah, Quebec, and Mexico.

Pyrite

Like chalcopyrite, pyrite is often called "fool's gold," because it is easily mistaken for the real thing. Finding pyrite is not a complete loss, however, as it often occurs with gold and silver. It is also used in the manufacture of sulfuric acid.

Identification Cannot be scratched by a knife; brittle; lacks cleavage; uneven fracture. Crystals isometric, usually as cubes and 12-sided crystals (called pyritohedrons); parallel striations on crystal faces. Pyrrhotite and marcasite lack the striations, have different crystal forms.

Colors Pale yellow to brass-yellow; easily tarnishes brown with film of iron oxide.

Environment Common in quartz veins, sometimes with gold, copper, and silver ore minerals; also in large masses in metamorphic rocks.

Occurrence Excellent crystals from Colorado, Nevada, Pennsylvania, and Utah.

Gold

Because of its rarity, gold is exceptionally valuable. Its name is said to come from an Anglo-Saxon word meaning "yellow." Gold starts out in quartz veins, and is released through a process of rock decay; it then becomes concentrated in the resulting sands, making it possible to "pan" for gold along streams and other alluvial and glacial deposits.

Identification Can be scratched by a knife; ductile, malleable; lacks cleavage; hackly fracture. Crystals isometric, usually 8-sided, 12-sided, and cubic; also as grains, lumps, leaves, wires. Gold-yellow color and malleability distinguish gold from copper and silver.

Colors Gold-yellow, brass-yellow, pale yellow; does not tarnish.

Environment Common in quartz veins with pyrite and arsenopyrite; also in deposits of sand and gravel.

Occurrence Famous gold mines in North America include those in Ontario, British Columbia, and Alaska; the Mother Lode belt in California; Colorado and Mexico.

Silver

Because of its malleable nature and occurrence as a natural metal, silver was one of the first metals to be worked by people. It often occurs with gold and mercury, and arsenic is also sometimes present. Silver is usually found in fairly small quantities.

Identification Can be scratched by a knife; malleable and ductile; lacks cleavage; hackly fracture. Crystals isometric, usually 8-sided, 12-sided, and cubic; also in grains, scales, micalike plates, or branching. Distinguished from lead on basis of color and hardness.

Colors Silver-white; tarnishes yellow, brown, black.

Environment Common in quartz veins in volcanic rocks; also associated with copper in metamorphosed basalt.

Occurrence Not common, but in several excellent localities in Arizona, Colorado, Michigan, Ontario, and Mexico.

Arsenopyrite

Formerly called mispickel, this mineral is an important ore of arsenic and is also mined for its gold content. The word arsenopyrite was coined as a contraction of "arsenical pyrites."

Identification
Cannot be scratched by a knife; brittle; one distinct cleavage; uneven fracture. Crystals monoclinic, prismatic and striated; also granular, compact, columnar. Crystal shape and color distinguish arsenopyrite from pyrite, marcasite, and pyrrhotite.

Colors
Silver-white to steel-gray.

Environment
Typically in veins associated with quartz, chalcopyrite, and gold; in granite pegmatites with quartz, muscovite, and microcline.

Occurrence
Widespread; most significant occurrences in California, Idaho, Maine, New Mexico, New York, Ontario, and Mexico.

Pectolite

Because of its compact fibrous nature the Greeks named this mineral *pektos*, meaning "compacted." Pectolite is closely related to rock-forming minerals of the pyroxene group, of which diopside is an important member. Pectolite has few industrial or commercial applications.

Identification Cannot be scratched by a knife; brittle; perfect cleavage in two directions at near right angles; splintery and uneven fracture; transparent; fluorescent. Crystals triclinic, rare; usually compact and radially fibrous in botryoidal masses. Radial structure and occurrence distinguish pectolite from tremolite and wollastonite.

Colors Colorless, white, grayish.

Environment Typically in basalt with calcite and amethystine quartz; occasionally in serpentinite.

Occurrence Excellent specimens from volcanic basalt in New Jersey; good crystals from Arkansas; and massive veins in California.

Pyrolusite

Often called manganese stain because it frequently occurs as a stain on manganese-bearing rocks, pyrolusite is the most common of several manganese oxide minerals. It is an important ore of manganese, and often forms as the result of alteration of another manganese-bearing mineral.

Identification
Cannot be scratched by a knife; brittle; one perfect cleavage; splintery and uneven fracture. Crystals tetragonal, rare; usually massive, columnar, fibrous, granular to powdery; forms branching growths on fracture surfaces. Difficult to distinguish from most other manganese oxide minerals.

Colors
Black to steel-gray; sometimes bluish where massive.

Environment
Alteration product of rhodochrosite; common in veins with calcite; occurs with barite and hematite.

Occurrence
Common in North America, but good pyrolusite fairly rare; known from Georgia and Minnesota.

Pyrophyllite

When heated, pyrophyllite peels in thin leaves—hence its name, from the Greek *pyr*, "fire," and *phyllon*, "leaf." It is used principally in ceramics and as a filler in paints and rubber products. Like talc, pyrophyllite is considered a soft mineral, and it easily reduces to a powder.

Identification   Easily scratched by a knife; greasy feel; perfect cleavage in one direction; uneven fracture; translucent to opaque. Crystals monoclinic, rare; commonly foliated, radially fibrous. Talc very similar but does not show fibrous radial structure.

Colors   White, apple-green, gray, yellow.

Environment   Principally in metamorphic rocks with quartz, albite, andalusite, and muscovite.

Occurrence   Fairly limited in distribution; excellent specimens in California, Georgia, and North Carolina.

Malachite

Easily cut and polished, malachite is the source of popular gemstones. Concentric bands of lighter and darker colors are a pleasing feature in massive malachite. Its Greek name comes from the leaf-green color. Malachite is also used as an ore of copper.

Identification | Can be scratched by a knife; brittle; one perfect cleavage crosswise; fracture curved, splintery; translucent. Crystals monoclinic, rare; usually short or long prismatic; also in botryoidal masses, radially fibrous, or stalactitic.

Colors | Emerald-green, grass-green, dark green.

Environment | An alteration product of copper-bearing sulfide minerals; found in copper deposits.

Occurrence | Excellent specimens occur in copper mines in the West, especially in Arizona, California, Nevada, and Utah; fine malachite also in Pennsylvania and Tennessee.

Actinolite

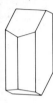

The name "actinolite" comes from the Greek *actinos*, meaning "ray," because this mineral commonly occurs in groups of radiating prismatic crystals. Actinolite exists in a dense, compact, and translucent form, called nephrite jade, that is used as a gemstone.

Identification Cannot be scratched by a knife; perfect cleavage in two directions lengthwise to form shape of a diamond; splintery fracture; transparent to translucent. Crystals monoclinic, usually long prisms; also radiating, columnar, divergent. Wollastonite similar but lacks prismatic habit and perfect cleavage.

Colors Bright to dark green, grayish-green, black.

Environment Metamorphic; develops in serpentinite and talc bodies; common in association with albite and muscovite in schists.

Occurrence Common to several types of rocks; finest specimens in serpentinite bodies in California and New England.

144

Talc

The basic ingredient in talcum powder, talc is widely used as a lubricant and as an insulation in electrical equipment. The name is derived from the Arabic word for mica, which is similar in certain regards. Massive talc is usually called soapstone.

Identification Easily scratched by a knife; smooth surface; greasy; in thin, flexible, micalike plates; perfect cleavage in one direction; translucent to opaque. Crystals monoclinic, rare; usually foliated, compact, massive. Softness and greasy feel distinguish talc from most other minerals; pyrophyllite usually not distinguishable in the field.

Colors Apple-green, white, pale buff.

Environment Usually in metamorphic rocks, such as talc schist, or in compact masses as soapstone; also in metamorphosed serpentinite and magnesium-rich limestone.

Occurrence Mined as soapstone in Connecticut, New York, and Vermont; pale green talc in New York (Staten Island) and Pennsylvania; dense talc in S. California.

146

Prehnite

Named for the Dutchman who introduced this mineral to Europe in 1774, prehnite is collected as a minor gemstone and is usually made into jewelers' cabochons. The best field mark is its botryoidal form—that is, in clusters that look like bunches of grapes.

Identification Cannot be scratched by a knife; brittle; cleavage good in one direction, poor in another; uneven fracture; translucent to almost transparent. Crystals orthorhombic, rare; more commonly seen in botryoidal masses; radially fibrous. Smithsonite similar but softer.

Colors Light green, gray, white, colorless.

Environment In cavities in volcanic basalt rocks; commonly with calcite and pectolite.

Occurrence Fairly common in Connecticut and Michigan; some fine prehnite from New Jersey and Virginia.

Smithsonite

James Smithson was a 19th-century mineralogist and patron of learning. Although English, he bequeathed a great deal of money for the establishment of a center of learning in this country, today known as the Smithsonian Institution. The mineral that bears his name is used as an ore of zinc; it is a secondary mineral, usually derived from sphalerite.

Identification   Can be scratched by a knife; brittle; perfect cleavage in three directions, forming a rhombohedron; fracture uneven to hackly. Crystals hexagonal, uncommon; commonly in botryoidal masses. Most carbonate minerals are softer; prehnite is harder.

Colors   White, gray, colorless, green, blue, yellow, pink, brown.

Environment   Develops where sphalerite has undergone alteration; frequently with galena, azurite, and malachite.

Occurrence   Arkansas, California, New Mexico, Pennsylvania, and Utah.

Opal

Because it lacks a crystalline structure, opal is generally not considered a mineral. It is a beautiful gem material and much sought throughout the world, although some people consider the stones tokens of bad luck.

Identification Cannot be scratched by a knife; brittle; lacks cleavage; breaks like glass, forming curved fractures. No crystals; massive. Curved fracture and opalescence (display of many colors) distinguishes opal from quartz.

Colors White, yellow, red, pink, brown, green, gray, blue; frequently displays many colors.

Environment In cavities and fractures in many kinds of rocks, most commonly in volcanic and sedimentary rocks.

Occurrence Best opal in W. United States, especially Oregon, Nevada, Utah, and Washington, and also in Mexico.

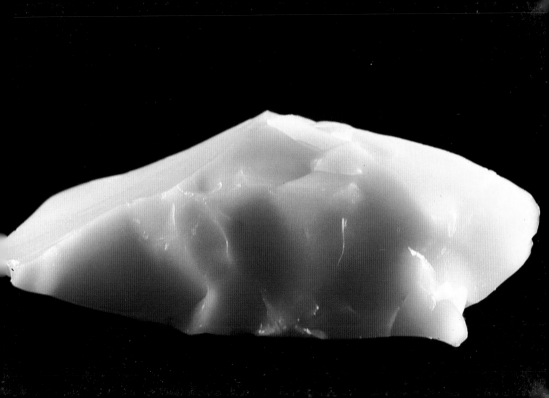

Wollastonite

A close relative of other rock-forming minerals in the pyroxene group, wollastonite is used as a filler in paints and ceramic materials. The name commemorates W. H. Wollaston, a British chemist and mineralogist.

Identification   Cannot be scratched by a knife; brittle; perfect cleavage in two directions at near right angles; fracture hackly and splintery; transparent to translucent; fluorescent. Crystals triclinic, usually elongated; also in fibrous masses, bladed. Tremolite has different cleavage angles; pectolite occurs with different associates.

Colors   Colorless, white, grayish; may have reddish or brownish tint.

Environment   Formed by metamorphism of impure limestones; masses of parallel or interlocking bladelike crystals are common; often associated with grossular and calcite as distinct bladelike crystals.

Occurrence   Masses are widespread; good crystals restricted to California, New Jersey, and New York.

Hematite

Named from the Greek word for blood, hematite often coats igneous and sedimentary rocks with a reddish color similar to rust, a form of hydrated iron oxide. An important ore of iron, hematite is also the source of red ochre, a pigment in paints.

Identification Cannot be scratched by a knife; brittle; lacks cleavage; uneven fracture, reflecting micalike plates. Crystals hexagonal, in thick to thin striated plates; also radiated, compact, botryoidal; earthy appearance. Its deep red powder distinguishes hematite from ilmenite and magnetite.

Colors Steel-gray, red, reddish-brown, black.

Environment Mined extensively from sedimentary rocks; also occurs commonly in igneous and metamorphic rocks.

Occurrence Abundant in iron mines in Michigan and Minnesota; nice crystals occur in New York; sharp black crystals in Arizona.

Dumortierite

Often mistaken for lapis lazuli, dumortierite is named in honor of Eugene Dumortier, a French paleontologist. It is easily cut and polished and is a source of gemstones. Although dumortierite is comparatively rare on this continent, a few sites in the West have yielded excellent specimens.

Identification Cannot be scratched by a knife; tough; good cleavage in one direction, poor in another; fracture uneven to hackly; transparent; sometimes fluorescent. Crystals orthorhombic, rare; usually in compact, fibrous, columnar masses. Resembles tourmaline but less bright with more fibrous appearance.

Colors Blue, violet; rarely pink or brown.

Environment With quartz and andalusite in granite pegmatites, and with andalusite and muscovite in metamorphic rocks.

Occurrence Best specimens from Arizona, Nevada, New York City, and scattered in California.

Azurite

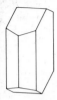

Easily cut and polished, azurite is a minor gemstone, but its main importance is as an ore of copper. Its intense color and hardness are good field marks. Often found with other copper-related minerals, especially malachite.

Identification Can be scratched by a knife; brittle; two good cleavages; fracture curved; transparent, in thin chips. Crystals monoclinic; commonly in well-formed tabular or equidimensional crystals; also radiating, in botryoidal masses, or incrusting.

Colors Dark blue.

Environment An alteration product of copper-bearing sulfide minerals, such as chalcopyrite; usually at shallow depths in earth's crust. Commonly associated with malachite and chalcopyrite.

Occurrence Best known from mines in W. North America, especially in Arizona, Utah, and Mexico.

Basalt

One of the most widespread igneous volcanic rocks, basalt occurs as massive lava flows covering vast areas over thousands of square miles. It is used as an ornamental stone and as an aggregate in concrete.

Identification   Fine-grained; may contain coarse crystals (phenocrysts) in fine-grained matrix. Composed of plagioclase feldspars and augite; may contain olivine or magnetite. May be dense or contain spherical cavities that represent bubbles of trapped gas.

Colors   Dark gray to black, dark brown to reddish-brown.

Environment   Forms when molten rock erupts on the surface of the earth or is intruded into fractures near earth's surface.

Occurrence   Hawaiian and Aleutian islands principally basalt; also large areas in West including Cascade Range in California, Oregon, and Washington; Columbia Plateau of Idaho, Oregon, and Washington; and mountains of Arizona, California, Nevada, New Mexico, Wyoming, and Mexico. Where volcanoes are and have been active.

Granite Pegmatite

Although not as common as granite, granite pegmatite is the source for gemstones of topaz, beryl, tourmaline, and garnets, and of many industrial minerals.

Identification Great variability of component minerals: grains range in size from a fraction of an inch to several feet. Light-colored minerals include quartz, muscovite, and potash feldspars; dark-colored minerals are tourmaline, garnet, and beryl. Many rare minerals also found in granite pegmatites.

Colors Usually white and speckled.

Environment With granite. Final stages of crystallization of molten rock containing volatiles rich in many chemicals and silica; all interact to produce different minerals.

Occurrence Generally in mountainous regions, especially in New England; also Appalachian Mountains, Black Hills of South Dakota, S. California, Colorado, Montana, and Ontario.

Granite

Because of its great strength and ability to take a high polish, granite is used widely in construction and as an ornamental stone for monuments. The terms "granite" and "granitic rocks" refer to several rocks that differ only slightly in chemical composition and are very similar to the untrained eye.

Identification Coarse-grained with grains mostly same size and easily seen. Speckled light and dark. Contains light-colored quartz and potash feldspars; also dark-colored biotite and hornblende.

Colors White to gray or dark gray, pink, or red.

Environment Forms slowly at great depths in earth's crust by crystallizing from molten rock. Brought to surface through mountain-building or exposed by deep erosion. Deposits of copper, gold, and tungsten often occur with granite.

Occurrence Mountains throughout E. and W. North America.

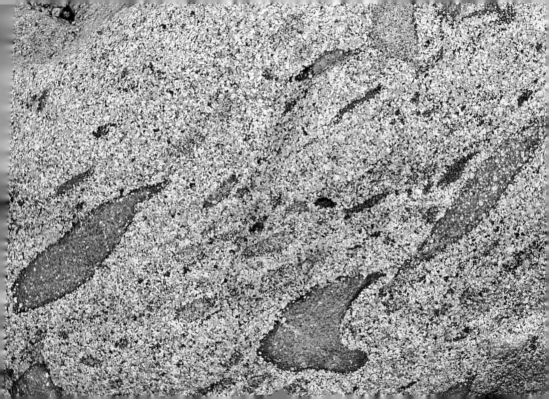

Limestone

Composed of fragments of coral and sea shells, limestone is mostly calcite. It is known for the great number of fossils it contains. Dense varieties are used in the construction of walls, the manufacture of concrete, and as a flux in iron- and steel-smelting operations; crushed limestone is employed as a concrete aggregate and road metal.

Identification Dense; grains even to uneven; fine-grained. Mostly calcite; may contain fossils; also quartz, clay, rock particles, and iron oxides. Can be scratched by a knife; generally reacts to weak acids by releasing gas bubbles. Often with shale and sandstone; looks pitted or crumbly.

Colors White, light gray to dark gray; also yellow and brown where iron oxides are present.

Environment Generally formed on ocean and sea floors where calcareous shell fragments deposit and become cemented.

Occurrence Hilly and mountainous regions of North America where soil cover is thin. Also underwater reefs and caves.

Shale

Because it is composed principally of clay, shale is used in the manufacture of pottery, brick, and many other ceramic products. On a small scale, it is used to produce a lightweight aggregate for concrete. Oil shale offers a great potential for fossil fuels. A massive, compact variety of shale is called mudstone.

Identification   Fine-grained. Particles are microscopic quartz and feldspars, and submicroscopic clay minerals; may have some organic matter; may contain fossils. Easily scratched by a knife. Well laminated; tends to split into flat, shell-like fragments parallel to laminations.

Colors   Usually light to dark gray; also buff, brown, reddish-brown, deep red.

Environment   Formed through compaction of very fine-grained sediments deposited in lake and ocean bottoms.

Occurrence   Widespread in North America; occurs where sandstone is found. Also abundant in coal and oil fields, but not necessarily exposed.

Sandstone

Firmly cemented massive sandstone is used for construction; it is easily carved and retains sculpted details for long periods of time. Red-brown sandstone, known as brownstone, has been used extensively for construction in the East. Deep sandstone deposits often contain petroleum; sandstone closer to the earth's surface is used in aquifers.

Identification Fine- to medium-grained; grains mostly rounded, sometimes angular. Predominantly quartz, with scarce feldspars, biotite, and magnetite; silica (quartz), calcite, clay, and iron oxides are cementing materials. Has gritty feel.

Colors Light to dark gray, red, brown, greenish.

Environment Formed through compaction and cementation of sandy sediments in beaches, deltas, floodplains, and deserts. Occurs in layers with other, more easily eroded sedimentary rocks; stands out against them.

Occurrence National parks in Arizona, Montana, and Utah; coast ranges of California and Oregon. Quarried in East.

Dolomite Rock

The terms "dolomite rock" and "dolomite" are often employed to mean a rock composed essentially of the mineral dolomite. Dolomite rock has many industrial uses, including as an aggregate in concrete and asphalt, in roofing granules, in iron smelting, and as a source of magnesium. It closely resembles limestone, from which it forms.

Identification Dense, rarely with grains of equal size; fine- to medium-grained; individual grains usually too small to be seen unaided; can be scratched by a knife. Often contains much calcite; also quartz and feldspars. Unlike limestone, rarely reacts to cold, weak acid solutions.

Colors White, light gray, tan.

Environment Formed on bottom of oceans and seas where limestone is acted upon by magnesium-rich water. Often has mineral veins and deposits of replacement minerals.

Occurrence Widespread throughout North America, except in Alaska, Delaware, Louisiana, New Hampshire, North Dakota, and Saskatchewan.

174

Schist

Although some schists are sufficiently stable to be used as building stone, the rock tends to split parallel to its inherent foliation and is not used in construction. Schist is sometimes used as flagstone and in the facing of exterior walls near ground level.

Identification
Uneven, granular, medium- to coarse-grained, with prominent parallel mineral orientation (schistosity). Light and dark speckles from constituent quartz, feldspars, muscovite, biotite, and hornblende. May contain almandine, staurolite, and kyanite.

Colors
Silvery white, shades of gray, yellowish, brownish.

Environment
Formed through metamorphism of sandstones and granite, utilizing much pressure.

Occurrence
Found in most places where slate and gneiss occur, especially in mountainous regions of California, New England, and British Columbia, as well as Adirondack Mountains, Appalachian Mountains, and Rocky Mountains.

Serpentinite

Often serpentinite is called serpentine; locally it is referred to as slickentite, because of the many polished, slippery surfaces of the rock. Used as a decorative stone, it is the source of chromite, platinum, and chrysotile asbestos. The soil where serpentinite is found is usually infertile and has scarce vegetation because of the high manganese content and thus lack of lime and alkalies.

Identification
: Dense, extremely fine-grained; grains cannot be seen by unaided eye; smooth to touch; can be scratched by a knife.

Colors
: Generally green, yellow-green, yellow to reddish-brown, dark green, black.

Environment
: Formed through metamorphism of magnesium-rich igneous plutonic rocks.

Occurrence
: Common in New England; also found in Maryland; underlies extensive areas in California, Oregon, and Washington.

Slate

Formerly quarried extensively for blackboards and roofing, slate is now used in the manufacture of roofing granules, clay pigeons for skeet shooting, and as flagstone. Slate metamorphoses from shale; it is susceptible to further metamorphism and likely to become, eventually, mica schist.

Identification Dense, fine-grained. Constituents of muscovite visible under a microscope. Slaty foliation produced by alignment of mica flakes in parallel planes, along which rock splits readily. Easily scratched by a knife.

Colors Medium to dark gray, black, green, brown, red, purple.

Environment Forms through metamorphism of shale or other fine-grained rocks. Pressure is essential in the process; in jagged outcrops, closely associated with schist and gneiss. Often with veins of quartz.

Occurrence Typical of mountainous regions of North America: California, Georgia, and Vermont.

Marble

All marble is metamorphosed limestone: Heat promotes uniform crystal size, and pressure helps to develop streakiness in some marbles. Because it takes a high polish, marble is used in statuary and for ornamental wall coverings. Impure varieties are used as concrete aggregate, road metal, and railroad ballast.

Identification   Even, granular, fine- to coarse-grained; grains can be seen by untrained eye. Pure marble consists mostly of calcite, but wollastonite and grossular may be present. Can be scratched by a knife; reacts to weak acids.

Colors   Pure marble white; impurities and minerals add green (talc and diopside), red (hematite), yellow (hydrated iron oxides), black (carbonaceous matter).

Environment   Formed when limestone is metamorphosed in earth's crust. Often with mica schists, gneisses, and intrusive igneous rocks.

Occurrence   Mountainous regions of North America. Quarried extensively in New England, Alabama, California, and Georgia.

Gneiss

This is a hard dense rock whose use in the building industry is limited to interior and exterior decorative purposes. The higher the mica content, the less useful gneiss becomes.

Identification Dense, hard, uneven, granular; medium- to coarse-grained; grains often seen by unaided eye, have more or less parallel orientation. Shows streaky, uneven, alternate layering of light-colored granular quartz and feldspars, and dark-colored biotite and hornblende. May contain almandine, kyanite, or staurolite.

Colors Too variable to be of diagnostic use, but alternating dark- and light-colored streaky layers are most characteristic of the rock.

Environment Formed through metamorphism using pressure of sandstone or granite plutonic rock.

Occurrence Mountainous regions underlain by old igneous plutonic and metamorphic rocks: New England, Adirondack Mountains, Appalachian Mountains, Canada, and Greenland.

Glossary

Amphibole
One of a group of closely related, dark-colored, rock-forming silicates, including actinolite and hornblende.

Asterism
The ability of some minerals to show a starlike figure in transmitted and reflected light.

Bedding
The arrangement of sedimentary rocks in approximately parallel layers (strata, or "beds"), corresponding to the original layers of deposits.

Borate
One of a group of boron-bearing minerals, such as colemanite.

Carbonaceous
Composed largely of organic carbon.

Clastic rocks
Sedimentary rocks consisting of fragments of other rocks and minerals that have been moved and redeposited to form a new rock, such as shale or sandstone.

Extrusive rock
Igneous rock that solidified on the surface of the earth.

Feldspars
A group of abundant rock-forming silicate minerals, including orthoclase and microcline, which belong to the potash feldspars, and albite, a member of the plagioclase, or soda-lime feldspars.

Foliated
Layered in leaflike pattern as seen in metamorphic rocks.

Igneous rock
Rock formed by the solidification of magma such as basalt.

Lamellar
Composed of thin layers, plates, or scales.

Magma
Molten rock material, beneath the solid crust of the earth, that solidifies to form igneous rocks at or below the earth's surface.

Metamorphic rock
Rock formed from preexisting rock

through heat, pressure, or chemical activity within the earth's crust, as in slate.

Mica
One of a group of silicate minerals that have perfect cleavage in one direction and can be easily split.

Organic compounds
Compounds produced in or by living organisms and containing carbon as the chief element.

Oxide
Any of a group of minerals in which oxygen combines with a metal, as in magnetite.

Pegmatite
An igneous rock containing extremely coarse grains.

Phenocryst
A prominent mineral surrounded by smaller mineral grains.

Pinacoid
An open form in which two opposing faces are the same.

Plutonic rock
A granular rock, such as granite, that solidified at great depth in the earth's crust.

Pyroxene
Any of a group of closely related, dark-colored, rock-forming silicate minerals, such as diopside, augite, and others.

Sedimentary rock
Layered rock formed through the solidification of sediments that originally consisted of mineral and rock grains or organic matter.

Silica
Another name for silicon dioxide, a particularly abundant mineral.

Silicate
Any of a huge group of minerals with silicon and oxygen as the essential elements. Silicates are the largest group of minerals.

Specular
Having the properties of a mirror.

Structure
Any large feature of a rock mass, such as bedding or layering.

Sulfate
A mineral composed of sulfur and oxygen, with other elements, such as gypsum.

Sulfide
A mineral composed of sulfur and one or more metals, such as pyrite or sphalerite.

Tungstate
A mineral composed of tungsten, oxygen, and another element, such as scheelite.

Ultrabasic rock
Any plutonic rock with very low silica content.

Vein
A tabular or sheetlike body of mineral matter that cuts across a rock; quartz, for example, often forms veins in granite or schist.

Vesicle
A small, somewhat spherical cavity in igneous volcanic rocks, often caused by bubbles of gas trapped in cooling magma.

Index

188

Photographers
All photographs were taken by Wolfgang Vogt with the exception of those listed below.

Rock H. Currier (85, 109)
Maurice Giles (59, 87)

Photo/Nats
Eugene H. Walker (167, 185)

Reo N. Pickens, Jr. (33, 139)

Root Resources
Louise K. Broman (61, 115, 125, 141, 161), Mary A. Root (31, 37, 41, 43, 49, 55, 81, 83, 101, 111, 121, 127, 131), Mary and Loren Root (63, 133)

Benjamin M. Schaub (89, 155)
Dr. Julius Weber (29, 35, 69, 71, 93, 105, 147, 153, 159)

Cover Photograph
Fluorite and dolomite by John Cancalosi

Title Page
Scheelite by
Dr. Julius Weber

Spread (24–25)
Opal by Martin Land/ Science Photo Library/ Photo Researchers, Inc.

Illustrators
Drawings by Judy F. L. Clinton and Ayn Svoboda